THIS BOOK BELONGS TO :

_____

NAME :

_____

FAVOURITE QUOTE :

_____

_____

CAT

Cats Live...

_____
_____
_____
_____
_____
_____
_____

Cats eat...

_____
_____
_____
_____
_____

Cats have...

_____
_____
_____

DOG

Dogs Live...

_____
_____
_____
_____
_____
_____
_____

Dogs eat...

_____
_____
_____
_____
_____

Dogs have...

_____
_____
_____

ELEPHANT

Elephants Live...

_____
_____
_____
_____
_____
_____
_____

Elephants eat...

_____
_____
_____
_____
_____

Elephants have...

_____
_____
_____

GIRAFFE

Giraffes Live...

_____
_____
_____
_____
_____
_____
_____

Giraffes eat...

_____
_____
_____
_____
_____

Giraffes have...

_____
_____
_____

TIGER

Tigers Live...

_____
_____
_____
_____
_____
_____
_____
_____

Tigers eat...

_____
_____
_____
_____
_____

Tigers have...

_____
_____
_____

PARROT

Parrots Live...

_____
_____
_____
_____
_____
_____
_____

Parrots eat...

_____
_____
_____
_____
_____
_____

Parrots have...

_____
_____
_____
_____

ZEBRA

Zebras Live...

_____
_____
_____
_____
_____
_____
_____

Zerbras eat...

_____
_____
_____
_____
_____

Zebras have...

_____
_____
_____
_____

WHALE

Whales Live...

_____
_____
_____
_____
_____
_____
_____
_____

Whales eat...

_____
_____
_____
_____
_____
_____

Whales have...

_____
_____
_____
_____

LION

Lions Live...

_____
_____
_____
_____
_____
_____
_____

Lions eat...

_____
_____
_____
_____
_____

Lions have...

_____
_____
_____

HEN

Hens Live...

_____
_____
_____
_____
_____
_____
_____

Hens eat...

_____
_____
_____
_____
_____

Hens have...

_____
_____
_____

DUCK

Ducks Live...
_____
_____
_____
_____
_____
_____
_____

Ducks eat...
_____
_____
_____
_____
_____

Ducks have...
_____
_____
_____

RABBIT

Rabbits Live...

_____
_____
_____
_____
_____
_____
_____

Rabbits eat...

_____
_____
_____
_____
_____

Rabbits have...

_____
_____
_____

COW

Cows Live...
_____
_____
_____
_____
_____
_____
_____
_____

Cows eat...
_____
_____
_____
_____
_____

Cows have...
_____
_____
_____

OSTRICH

Ostriches Live...
_____
_____
_____
_____
_____
_____
_____

Ostriches eat...
_____
_____
_____
_____
_____

Ostriches survive...
_____
_____
_____

PANDA

Pandas Live...

_____
_____
_____
_____
_____
_____
_____

Pandas eat...

_____
_____
_____
_____
_____

Pandas have...

_____
_____
_____

GORILLA

Gorillas Live...

_____
_____
_____
_____
_____
_____
_____

Gorillas eat...

_____
_____
_____
_____
_____

Gorillas survive...

_____
_____
_____

LEMUR

Lemurs Live...

_____
_____
_____
_____
_____
_____
_____

Lemurs eat...

_____
_____
_____
_____

Lemurs have...

_____
_____
_____

ALLIGATOR

Alligators Live...
_____
_____
_____
_____
_____
_____
_____

Alligators eat...
_____
_____
_____
_____
_____

Alligators have...
_____
_____
_____

ANT

Ants Live...
_____
_____
_____
_____
_____
_____
_____
_____

Ants eat...
_____
_____
_____
_____
_____

Ants have...
_____
_____
_____
_____

BUTTERFLY

Butterflies Live...

_____
_____
_____
_____
_____
_____
_____
_____

Butterflies eat...

_____
_____
_____
_____
_____

Butterflies have...

_____
_____
_____

FISH

Fish Live...

_____
_____
_____
_____
_____
_____
_____

Fish eat...

_____
_____
_____
_____
_____

Fish have...

_____
_____
_____

RAT

Rats Live...

_____
_____
_____
_____
_____
_____
_____

Rats eat...

_____
_____
_____
_____
_____

Rats survive...

_____
_____
_____

TORTOISE

Tortoises Live...

_____
_____
_____
_____
_____
_____
_____

Tortoises eat...

_____
_____
_____
_____
_____

Tortoises have...

_____
_____
_____

GUINEA PIG

Guinea pigs Live…
_____
_____
_____
_____
_____
_____
_____
_____

Guinea pigs eat…
_____
_____
_____
_____
_____

Guinea pigs have…
_____
_____
_____

TAPIR

Tapirs Live...
_____
_____
_____
_____
_____
_____
_____

Tapirs eat...
_____
_____
_____
_____
_____

Tapirs have...
_____
_____
_____

www.ingramcontent.com/pod-product-compliance
Lightning Source LLC
Chambersburg PA
CBHW080908220526
45466CB00011BA/3514